Lecture Notes in Mathematics

A collection of informal reports and seminars
Edited by A. Dold, Heidelberg and B. Eckmann, Zürich

93

K. R. Parthasarathy

University of Manchester, Manchester, England

Multipliers on Locally Compact Groups

1969

Springer-Verlag Berlin · Heidelberg · New York

Library of Congress Catalog Card Number 71-84142 Printed in Germany. Title No. 3699

C O N T E N T S

MULTIPLIERS IN LOCALLY COMPACT GROUPS

1. INTRODUCTION

In the mathematical formulation of quantum mechanics proposed by Von Neumann (cf [9], [14], [15]) the set of all propositions concerning a quantum mechanical system is an ortho-complemented lattice. The simplest example of such a lattice is the lattice $\mathcal{L}(\mathfrak{h})$ of all closed subspaces of a separable Hilbert space $\mathfrak{h}$. The observables in such a system turn out to be self adjoint operators in $\mathfrak{h}$. In order to construct the standard observables like energy, linear, angular and spin angular momenta, etc., in such a system it is necessary to study the effect of coordinate transformations by a group G of symmetries. In this context the study of representations of G in the group of automorphisms of the lattice $\mathcal{L}(\mathfrak{h})$ is of great importance.

By a theorem of Wigner (cf [14] Theorem 7.27, page 167) it is known that every automorphism τ of $\mathcal{L}(\mathfrak{h})$ is induced by a unitary or antiunitary operator U^{τ}. The operator U^{τ} is determined uniquely upto a scalar multiple of modulus unity.

Let $g \to \tau_g$ be a representation of G in the group of automorphisms of $\mathcal{L}(\mathfrak{h})$. Suppose τ_g is induced by the operator U_g which may be unitary or auniunitary. If the group G is sufficiently regular (as for example a connected Lie group) we can take U_g to be unitary for all $g \varepsilon G$. Since τ_e is the identity automorphism we can take U_e to be the identity operator. Since $U_{g_1} U_{g_2}$ and $U_{g_1 g_2}$ induce the same automorphism $\tau_{g_1 g_2}$ it follows that there exists a constant $\sigma(g_1, g_2)$ of modulus unity such that

$$U_{g_1} U_{g_2} = \sigma(g_1, g_2) U_{g_1 g_2} \quad \text{for all} \quad g_1, g_2 \varepsilon G.$$

Since

$$U_{g_1} (U_{g_2} U_{g_3}) = (U_{g_1} U_{g_2}) U_{g_3} \quad \text{for all } g_1, g_2, g_3 \varepsilon G,$$

we have

$$\sigma(g_1, g_2 g_3) \sigma(g_2, g_3) = \sigma(g_1, g_2) \sigma(g_1 g_2, g_3) \quad \text{for all } g_1, g_2, g_3 \varepsilon G \tag{1.1}$$

Since $U_e = I$,

$$\sigma(g, e) = \sigma(e, g) = 1 \quad \text{for all} \quad g \varepsilon G. \tag{1.2}$$

Any function σ defined on G x G, taking values in the multiplicative group of all complex numbers of modulus unity and satisfying equations (1.1) and (1.2) is called a multiplier on G x G.

The purpose of this article is to make a systematic analysis of the multipliers on certain locally compact groups. Such an analysis is, of course, not new. It was originally done by Bargmann

[1], and then by Mackey [8], [10], Varadarajan [13], Simms [12] and many others. The presentation given here is probably more unified and self-contained, many of the proofs are simpler and some of the results are more general. The main ideas are contained in [1], [8], [10] and [13].

Keeping the mathematical physicist readers in mind every effort has been made to keep the presentation as elementary and self-contained as possible. However, a knowledge of Borel structures and measures in metric spaces and the basic theory of Lie groups and Lie algebras has been assumed. Regarding these the reader may refer to [7], [2], [4] and [5] respectively.

2. STANDARD GROUPS WITH A RIGHT INVARIANT MEASURE

Let G be a group and $\mathcal{B}$ be a σ-field of subsets of G such that the mapping $(x,y)\to xy^{-1}$ from G x G onto G is measurable. Here G x G is given the product Borel structure. $(G, \mathcal{B})$ is called a <u>measurable group</u>. Suppose there exists a complete and separable metric space X such that there is a one-one map γ from G onto X where γ and γ^{-1} are both measurable and X is given its usual Borel structure. Then $(G, \mathcal{B})$ is called a <u>standard group</u>.

Throughout our discussions we shall denote by e the identity element of G. For any set $E\subset G$ and any element $x\in G$ we define the sets E^{-1}, Ex and xE as follows: $E^{-1} = \{y: y^{-1}\in E\}$, $Ex = \{yx: y\in E\}$, $xE = \{xy: y\in E\}$. If E_1 and E_2 are two subsets of G we denote by

E_1E_2 the set $\{xy: x\epsilon E_1, y\epsilon E_2\}$.

A measure μ defined on the measurable group $(G, \mathcal{B})$ is said to be right invariant if $\mu(E) = \mu(Ex)$ for all $E\epsilon \mathcal{B}$ and $x\epsilon G$. It is clear that $Ex\epsilon \mathcal{B}$ whenever $E\epsilon \mathcal{B}$. Left invariance is defined in a similar way.

We shall now analyse some of the properties of a standard group $(G, \mathcal{B})$ with a non trivial right invariant σ-finite measure μ. We follow Mackey [7]. Throughout this section we shall assume that $(G, \mathcal{B}, \mu)$ is a standard group with a σ-finite right invariant measure μ.

Consider the complex Hilbert space $\mathcal{L}_2(\mu)$ of all functions square integrable with respect to μ. $\mathcal{L}_2(\mu)$ is separable since $\mathcal{B}$ is countably generated. Let $\mathcal{U}$ be the space of unitary operators on $\mathcal{L}_2(\mu)$, endowed with the weak topology. Let $\mathcal{B}_{\mathcal{U}}$ be the smallest σ-field generated by the open subsets of $\mathcal{U}$. Then $(\mathcal{U}, \mathcal{B}_{\mathcal{U}})$ is a standard group.

We shall now introduce a mapping from $(G, \mathcal{B})$ into $(\mathcal{U}, \mathcal{B}_{\mathcal{U}})$ as follows: for every $x\epsilon G$ define the unitary operator R_x by

$$(R_x f)(y) = f(yx) \quad \text{for all } f\epsilon \mathcal{L}_2(\mu). \tag{2.1}$$

The right invariance of μ implies that R_x is unitary. By considering the product measure space $(G \times G, \mathcal{B} \times \mathcal{B}, \mu \times \mu)$ and applying Fubini's theorem (cf [2]) we conclude that the map $x \to R_x$ is measurable. Further $R_x R_y = R_{xy}$ and $R_e = I$, where I denotes the identity operator.

Lemma 2.1 The mapping $x \to R_x$ is one-one.

Proof: Suppose for two points x and y, $R_x = R_y$. Then $R_{xy^{-1}} = R_x R_y^{-1} = I$. Thus it is enough to show that for any $x_o \varepsilon G$, $R_{x_o} = I$ implies that $x_o = e$. If $R_{x_o} = I$, we have from (2.1)

$$\chi_E (x\, x_o) = \chi_E (x) \quad \text{a.e.}(\mu) \qquad \text{for all } E \varepsilon \mathcal{B} ,$$

where χ_E stands for the characteristic function of the set E. Hence

$$\mu(E\, x_o^{-1} \Delta\, E) = \int |\chi_E (x\, x_o) - \chi_E (x)| d\mu(x) = 0 \quad \text{for all } E \varepsilon \mathcal{B} , \qquad (2.2)$$

where Δ stands for symmetric difference. Since G has a separable metric topology whose Borel sets constitute $\mathcal{B}$, we can choose a metric d in G and a dense sequence $z_1, z_2, \ldots$ in this metric such that every non empty sphere is an element of $\mathcal{B}$ and contains a member of the sequence $\{z_n\}$.

Let $S_{jr} = \{x: d(x, z_j) < r\}$ for every positive rational number r. Then from (2.2) we have

$$\mu(S_{jr}\, x_o^{-1} \Delta\, S_{jr}) = 0 \qquad \text{for all j and r.}$$

Thus

$$\mu\Big\{\bigcup_{j,r} ([x: d(x\, x_o, z_j) < r, \;\; d(x, z_j) \geq r] \cup$$
$$[x: d(x\, x_o, z_j) \geq r, \;\; d(x, z_j) < r \;\;)]\Big\} = 0.$$

The set appearing within the wavy brackets above is nothing but

$\{x: x\,x_o \neq x\}$. Thus

$$\mu\,\{x: x\,x_o \neq x\} = 0$$

Hence $x_o = e$. This completes the proof.

<u>Lemma 2.2</u> The range of the map $x \to R_x$ is a Borel subset of $\mathcal{U}$.

<u>Proof</u>: Since $x \to R_x$ is a one-one measurable map of a standard space into another this lemma is an immediate consequence of Kuratowski's theorem (cf [11], page 21).

We shall now assign the space G the smallest topology which makes the map $x \to R_x$ continuous. This is called the <u>Weil topology</u> of G with respect to μ. From Lemmas 2.1 and 2.2 it is clear that the Borel σ-field generated by the open subsets of the Weil topology in G is the same as $\mathcal{B}$. Further G becomes a topological group in the Weil topology. We shall soon establish that this topology is actually locally compact. To this end we need a lemma.

<u>Lemma 2.3</u> Let $(G, \mathcal{B})$ be a standard group with a non trivial right invariant measure μ. Then for any $\delta>0$ and any $E \in \mathcal{B}$ such that $\mu(E)<\infty$ the set

$$\{x: \mu(Ex \,\Delta\, E) < \delta\}$$

is open in the Weil topology.

<u>Proof</u>: This follows immediately from the fact that the map

$$x \to \mu(Ex \,\Delta\, E) = \int |(R_x \chi_E)(y) - \chi_E(y)|^2 \, d\mu(y)$$

is continuous in the Weil topology.

<u>Theorem 2.1</u> (G. W. Mackey [7])

Let $(G, \mathcal{B})$ be a standard group with a non trivial σ-finite right invariant measure μ. Then the Weil topology of G with respect to μ is locally compact.

<u>Proof:</u> In order to prove the theorem it is enough to exhibit a compact set with a non empty interior. Since the Weil topology makes G a standard topological group it follows that there exists a compact set K in this topology such that $0 < \mu(K) < \infty$ (cf [11], page 29). Consider the set

$$K_1 = \{x: \mu(Kx \,\Delta\, K) < \delta\}$$

where $\delta = \mu(K)$. By Lemma 2.3, K_1 is open. Since $e \in K_1$, K_1 is non-empty. We shall now complete the proof by showing that the compact set $K^{-1}K$ contains K_1. Indeed $x \in K_1$ implies that

$$\mu(Kx \cap K') + \mu(K'x \cap K) < \delta$$

where "'" denotes complementation. Hence, either

$$\mu(Kx \cap K') < \delta/2,$$

or

$$\mu(K'x \cap K) < \delta/2.$$

In the first case

$$\delta = \mu(K) = \mu(Kx) = \mu(Kx \cap K') + \mu(Kx \cap K) < \delta/2 + \mu(Kx \cap K).$$

Hence

$$\mu(Kx \cap K) > \delta/2.$$

Thus $Kx \cap K \neq \emptyset$ and $x \varepsilon K^{-1}K$. In the second case

$$\delta = \mu(K) = \mu(K \cap K'x) + \mu(K \cap Kx) < \delta/2 + \mu(K \cap Kx).$$

By the same kind of argument we conclude that $x \varepsilon K^{-1}K$. Thus $K_1 \subset K^{-1} K$. This completes the proof.

<u>Corollary 1</u> If μ is a totally finite nontrivial right invariant measure on a standard group $(G, \mathcal{B})$ then the Weil topology is compact.

<u>Proof:</u> Without loss of generality we may assume that $\mu(G)=1$. By the above theorem G is locally compact in the Weil topology. Hence the Haar measure is unique. Since $\int \mu(Ex^{-1})\, d\mu(x) = \mu(E)$ for all $E \varepsilon \mathcal{B}$, it follows by the argument in pages 61-63 of [11] that G is compact.

We shall now prove a result concerning homomorphisms of locally compact groups which we shall have occasion to use later. To this end we need the following lemma.

<u>Lemma 2.4</u> Let G be a locally compact second countable group and μ be its right invariant Haar measure. Let E be any Borel set such that $0 < \mu(E) < \infty$. Then $E^{-1}E$ contains a

neighbourhood of e.

Proof: Since $\mu(E) = \sup\{\mu(K),\ K \subseteq E, K \text{ compact}\}$ it is enough to prove the lemma when E itself is compact. Suppose E is compact. Then exactly as in the proof of Theorem 2.1,

$$E^{-1}E \supseteq \{x: \mu(Ex \Delta E) < \mu(E)\} .$$

Thus it is enough to show that the right hand side of the above inclusion relation has a non empty interior. Since μ is regular, for any $\varepsilon>0$, we can find an open set $U \supseteq E$ such that

$$\mu(U - E) < \varepsilon/2 . \tag{2.3}$$

Since E is compact we can find a symmetric open set V containing e such that

$$E\ V \subseteq U \tag{2.4}$$

Indeed if there does not exist any such V we can find a sequence $x_n \to e$ and a sequence $y_n \varepsilon E$ such that $y_n x_n \varepsilon U'$ for all n. Since E is compact we may assume that y_n is convergent. Suppose $\lim_n y_n = y$. Since U' is closed, $y = \lim y_n x_n \varepsilon U'$. But $y \varepsilon E \subseteq U$. This is a contradiction. Hence (2.4) holds. From (2.3) and (2.4) we have for all $x \varepsilon V$

$$\begin{aligned}
\mu(Ex \Delta E) &= \mu(Ex \cap E') + \mu(E \cap E'x)\\
&\leq \mu(E V \cap E') + \mu(Ex^{-1} \cap E')\\
&\leq 2\mu(E\ V \cap E')\\
&\leq 2\mu(U \cap E')\\
&\leq \varepsilon.
\end{aligned}$$

If we choose $\varepsilon=\mu(E)$, we have

$$V \subseteq \{x: \mu(Ex \,\Delta\, E) < \mu(E)\}.$$

Since V is open this completes the proof of the lemma.

Theorem 2.2 (Mackey [7])

Let G be a locally compact second countable group and H a separable metric group. Let $\pi: G \to H$ be a Borel measure homomorphism from G into H. Then π is continuous.

Proof: Without loss of generality we may assume that π is onto. Let μ be the right invariant Haar measure of G. Let V be any open neighbourhood of the identity in H and $V_1 \subseteq V$ be a symmetric neighbourhood of the identity such that $V_1.V_1 \subseteq V$. The theorem will be proved if $\pi^{-1}(V)$ contains a neighbourhood of the identity in G. Let $h_1, h_2, \ldots, h_n, \ldots$ be an everywhere dense sequence in H. Then

$$H = \bigcup_n V_1\, h_n.$$

Hence

$$G = \pi^{-1}(H) = \bigcup_n \pi^{-1}(V_1\, h_n).$$

Let g_n be so chosen that $\pi(g_n)=h_n$. Then

$$G = \bigcup_n \pi^{-1}(V_1)\, g_n.$$

Hence for some n, $\mu(\pi^{-1}(V_1)g_n)>0$. Since μ is right invariant,

$\mu(\pi^{-1}(V_1))>0$. Further

$$\pi^{-1}(V_1)\ \pi^{-1}(V_1) \subseteq \pi^{-1}(V_1.V_1) \subseteq \pi^{-1}(V). \qquad (2.5)$$

Since V_1 is symmetric $\pi^{-1}(V_1)$ is also symmetric. Since μ is σ-finite we can choose a compact set K such that $K \subsetneq \pi^{-1}(V_1)$ and $0<\mu(K)<\infty$. From (2.5) we have $K^{-1}K \subseteq \pi^{-1}(V)$. Hence by Lemma 2.4, $\pi^{-1}(V)$ includes a neighbourhood of e. This completes the proof.

Corollary 1 Let G be a locally compact second countable group with μ as the right invariant Haar measure. Its topology coincides with the Weil topology.

Proof: Let us denote the abstract group G with its Weil topology by $\tilde{G}$. Consider the identity map $x \to x$ from G to $\tilde{G}$. This is a Borel isomorphism. Hence by Theorem 2.2, it is a homeomorphic isomorphism. This completes the proof.

3. BOREL MULTIPLIERS ON A LOCALLY COMPACT GROUP

Throughout this section we shall denote by G a locally compact second countable group. Let T denote the multiplicative group of all complex numbers of modulus unity.

Definition 3.1 A function σ: G x G $\to$ T is called a Borel multiplier if σ is a Borel measurable and satisfies equations (1.1) and (1.2).

Since we are going to deal only with Borel multipliers in what follows we shall call any Borel multiplier simply a multiplier. A multiplier σ is said to be trivial if there exists a Borel function $a: G \to T$ such that

$$\sigma(g_1,g_2) = a(g_1)\, a(g_2)\, a(g_1\, g_2)^{-1} \quad \text{for all } g_1,g_2 \in G.$$

It is clear that if σ_1 and σ_2 are multipliers then their product $\sigma_1\sigma_2$ is also a multiplier. If σ is a multiplier, σ^{-1} is also a multiplier. Thus all the multipliers constitute an abelian group $M(G)$. Two multipliers σ_1 and σ_2 are said to be equivalent ($\sigma_1 \sim \sigma_2$ in symbols) if $\sigma_1\sigma_2^{-1}$ is trivial. It is clear that "$\sim$" is indeed an equivalence relation. It is obvious that all the trivial multipliers constitute a subgroup $M_o(G)$. The quotient group $M(G)/M_o(G)$ is the set of all equivalence classes of multipliers.

Two multipliers σ_1 and σ_2 are said to be locally equivalent ($\sigma_1 \overset{loc}{\sim} \sigma_2$ in symbols) if there exists a neighbourhood N of e in G and a Borel function $a:N \to T$ such that $(\sigma_1\sigma_2^{-1})(g_1,g_2) = a(g_1)\, a(g_2)\, a(g_1g_2)^{-1}$ for all $g_1,g_2 \in N$. In particular a multiplier σ is said to be locally trivial if there exists a neighbourhood N of e in G and a Borel function $a:N \to T$ such that $\sigma(g_1,g_2) = a(g_1)\, a(g_2)\, a(g_1g_2)^{-1}$ for all $g_1,g_2 \in N$.

Given a multiplier σ, following Mackey (cf[8]) we shall construct a group G^σ called the σ extension of G as follows: The set G^σ is the same as the cartesian product $G \times T$; the group operation 'o' is defined by

$$(g_1, \lambda_1) \circ (g_2, \lambda_2) = (g_1 g_2, \lambda_1 \lambda_2 \sigma(g_1, g_2)).$$

The element (e, 1) where e is the identity of G is called the identity of G^σ.

Simple calculation shows that 'o' is associative and the inverse of (g, λ) is uniquely defined and is equal to $(g^{-1}, \lambda^{-1} \sigma(g, g^{-1})^{-1})$ for all $(g, \lambda)\epsilon G^\sigma$. Thus G^σ is a group. We give G^σ the product Borel structure of G x T. Since σ is measurable on G x G it is clear that G^σ becomes a standard group (cf §2).

We shall denote by $d\mu$ and $d\lambda$ under integral signs to denote integration with respect to the right invariant Haar measure μ in G and the normalised uniform measure in T. It follows from Fubini's theorem that for any bounded measurable function $f(g, \lambda)$ on G^σ,

$$\iint f(g\, g_o, \lambda\, \lambda_o\, \sigma(g, g_o))\, d\mu(g) d\lambda = \iint f(g, \lambda)\, d\mu(g) d\lambda$$

$$\text{for all } (g_o, \lambda_o)\epsilon G^\sigma.$$

This shows that the product of μ in G and the uniform measure in T is right invariant in G^σ. Thus G^σ becomes a standard group with a σ-finite right invariant measure. We shall hereafter consider the topological group G^σ with its Weil topology. By Theorem 2.1 G^σ becomes a locally compact topological group. Let $T^\sigma = \{(e,\lambda):\lambda\epsilon T\}$. T^σ is a topological subgroup of G^σ which is standard and has an invariant measure. Further T^σ is Borel isomorphic with T and hence by Theorem 2.2 is homeomorphically isomorphic with T. Thus T^σ is a compact subgroup of G^{σ}. Since $(g_o,\lambda_o)o(e,\lambda) = (g_o, \lambda_o\lambda) = (e,\lambda)o(g_o,\lambda_o)$ for all $\lambda\varepsilon T$, $(g_o,\lambda_o)\varepsilon G^{\sigma}$, it follows that T^{σ} is in the centre of G^{σ} and T^{σ} is a normal subgroup of G^{σ}. We shall denote by τ the homomorphism

$$\tau: (g, \lambda) \to g \tag{3.1}$$

from G^{σ} onto G. The kernel of τ is precisely T^{σ}. If we endow G with the quotient topology, then it becomes a locally compact group with the same Borel structure as that of its original topology. Hence by Theorem 2.2 the quotient topology coincides with the original topology of G. The following lemma puts these remarks together.

<u>Lemma 3.1</u> The group G^{σ} with its Weil topology is a locally compact topological group. The subset $T^{\sigma}=\{(e,\lambda),\lambda\varepsilon T\}$ is a compact subgroup of G^{σ} which is contained in the centre of G^{σ}. The quotient group G^{σ}/T^{σ} is isomorphic with G.

<u>Proof:</u> For every $(g,\lambda)\varepsilon G^{\sigma}$, we define the unitary operators U_g and $U_{(g,\lambda)}$ in $\mathcal{L}_2(\mu)$ as follows:

$$\left.\begin{aligned} (U_g f)(h) &= \sigma(h, g)\, f(hg) \\ (U_{(g,\lambda)} f)(h) &= \lambda\sigma(h, g)\, f(hg) \end{aligned}\right\} \quad \text{for all } f\varepsilon\mathcal{L}_2(\mu) \tag{3.2}$$

It follows from the definition of a multiplier that

$$U_{g_1} U_{g_2} = \sigma(g_1,g_2)\, U_{g_1 g_2} \qquad \text{for all } g_1,g_2 \varepsilon G$$
$$U_{(g_1,\lambda_1)} U_{(g_2,\lambda_2)} = U_{(g_1 g_2,\ \lambda_1\lambda_2\ \sigma(g_1,g_2))} \quad \text{for all } (g_1,\lambda_1),(g_2,\lambda_2) \varepsilon G^{\sigma} \tag{3.3}$$

<u>Lemma 3.2</u> The mapping $(g, \lambda) \to U_{(g,\lambda)}$ is a weakly continuous unitary representation of G^{σ} in $\mathcal{L}_2(\mu)$.

<u>Proof:</u> From (3.2) and (3.3) it is clear that $(g, \lambda) \to U_{(g, \lambda)}$ is a Borel representation of the locally compact group G^{σ} into the standard group $\mathcal{U}$ of all unitary operators on $\mathcal{L}_2(\mu)$. Hence by Theorem 2.2, this representation is weakly continuous. This completes the proof of the lemma.

<u>**Theorem 3.1**</u> (Mackey [8])

Let G be a locally compact second countable group and σ be a multiplier on G x G. Then there exists an equivalent multiplier σ' such that σ' is continuous on N x N where N is a neighbourhood of e in G.

<u>Proof:</u> Define the operators U_g and $U_{(g,\lambda)}$ as in (3.2) for every $(g,\lambda) \varepsilon G^{\sigma}$. Let us denote the inner product in $\mathcal{L}_2(\mu)$ by $\langle \cdot,\cdot \rangle$. From (3.2) we have for any $f \varepsilon \mathcal{L}_2(\mu)$

$$| \langle U_g f, f \rangle | = | \langle U_{(g, \lambda)} f, f \rangle | . \tag{3.4}$$

By Lemmas 3.1 and 3.2 it is clear that $|\langle U_g f,f \rangle|$ is a continuous function of g for every $f \in \mathcal{L}_2(\mu)$. Choose and fix an $f_o \in \mathcal{L}_2(\mu)$, $f_o \neq 0$. Then we have

$$\lim_{g \to e} |\langle U_g f_o, f_o \rangle| = |\langle f_o, f_o \rangle| \neq 0.$$

Hence there exists a neighbourhood N of e such that

$$\langle U_g f_o, f_o \rangle \neq 0 \qquad \text{for all } g \in N.$$

Define the Borel function

$$a(g) = \begin{cases} \dfrac{\langle U_g f_o, f_o \rangle}{|\langle U_g f_o, f_o \rangle|} & \text{if } g \in N \\ 1 & \text{if } g \notin N \end{cases} \tag{3.5}$$

Then $|a(g)| = 1$. Define the unitary operator V_g by putting

$$V_g = a(g)^{-1} U_g. \tag{3.6}$$

We have from (3.3)

$$\begin{aligned} V_{g_1} V_{g_2} &= a(g_1)^{-1} a(g_2)^{-1} U_{g_1} U_{g_2} \\ &= \sigma(g_1,g_2)\, a(g_1)^{-1} a(g_2)^{-1} a(g_1 g_2)\, V_{g_1 g_2} \\ &= \sigma'(g_1,g_2)\, V_{g_1 g_2} \qquad \text{for all } g_1, g_2 \in G \end{aligned} \tag{3.7}$$

where

$$\sigma'(g_1,g_2) = \sigma(g_1,g_2)\, a(g_1)^{-1} a(g_2)^{-1} a(g_2)^{-1} a(g_1 g_2) \tag{3.8}$$

In particular σ' is a multiplier on G x G which is equivalent to σ. Consider now the group $G^{\sigma'}$ and the unitary operator $V_{(g,\lambda)}$ on $\mathcal{L}_2(\mu)$ defined by

$$V_{(g,\lambda)} f = \lambda V_g f. \tag{3.9}$$

Equations (3.7) and (3.9) imply that $(g,\lambda) \to V_{(g,\lambda)}$ is a Borel representation of $G^{\sigma'}$. Hence by Theorem 2.2, it is a weakly continuous representation of $G^{\sigma'}$. From (3.5), (3.6) and (3.9) we have

$$|\langle V_{(g,\lambda)} f_o, f_o \rangle| = |\langle V_g f_o, f_o\rangle| \quad \text{for all } (g,\lambda)\varepsilon G^{\sigma'},$$

$$\langle V_g f_o, f_o\rangle \quad = |\langle U_g f_o, f_o\rangle| > 0 \quad \text{for all } g\varepsilon N$$

Thus

$$\langle V_g f_o, f_o\rangle \quad = |\langle V_g f_o, f_o\rangle| > 0 \quad \text{for all } g\varepsilon N.$$

Hence the two maps

$$(g, \lambda) \to \lambda \langle V_g f_o, f_o\rangle$$

$$(g, \lambda) \to \langle V_g f_o, f_o\rangle$$

are continuous in $\tau'^{-1}(N)$ where τ' is the map $(g,\lambda) \to g$ from $G^{\sigma'}$ onto G. Since $\langle V_g f_o, f_o\rangle > 0$ in N, it follows that the map

$$(g, \lambda) \longrightarrow \lambda$$

is continuous on $\tau'^{-1}(N) = \tilde{N}$ say. Choose a neighbourhood M of the identity in $G^{\sigma'}$ such that $M M \subseteq \tilde{N}$. Then the maps

$$(g_1, \lambda_1), (g_2, \lambda_2) \rightarrow \lambda_1 \lambda_2$$

$$(g_1, \lambda_1), (g_2, \lambda_2) \rightarrow \lambda_1 \lambda_2 \sigma' (g_1, g_2)$$

are continuous on M x M. Hence the map

$$((g_1, \lambda_1), (g_2, \lambda_2)) \rightarrow \sigma' (g_1, g_2)$$

is continuous on M x M. Let $\tau'(M) = N_1$. Then by Lemma 3.1 the map

$$(g_1, g_2) \rightarrow \sigma' (g_1, g_2)$$

is continuous on $N_1 \times N_1$ and N_1 is a neighbourhood of e in G. This completes the proof.

For later use we need the following two lemmas.

Lemma 3.3 Let G be a locally compact second countable group and σ be a multiplier on G x G. Suppose there exists a neighbourhood N of e such that σ is continuous in N x N. Then there is a neighbourhood of the identity in G^σ where the map $(g, \lambda) \rightarrow \lambda$ from G^σ into T is continuous.

Proof: Consider the operators U_g and $U_{(g,\lambda)}$ defined by (3.2) and (3.3). If f_o is a non zero function in $\mathcal{L}_2(\mu)$ vanishing outside N, then it is clear from (3.2) that $\langle U_g f_o, f_o \rangle$ is continuous in N. From the proof of Theorem 3.1 it is clear that there exists a neighbourhood N_1 of e such that $N_1 \subseteq N$ and $\langle U_g f_o, f_o \rangle \neq 0$ for all $g \in N_1$. Further the map

$$(g, \lambda) \to \lambda \langle U_g f_o, f_o \rangle$$

is continuous in G^σ. By Lemma 3.1 it follows that the map $(g, \lambda) \to \lambda$ is continuous in $\tau^{-1}(N_1)$ where τ is defined by (3.1). This completes the proof.

Remark If σ is continuous in a neighbourhood of the identity in G x G, it follows from the above lemma and the local compactness of G^σ that there exists a neighbourhood of the identity in G^σ such that the identity map on this neighbourhood is a homeomorphism into the product topological space G x T.

Lemma 3.4 Let G be a locally compact second countable group and σ be a trivial multiplier on G x G. Suppose there exists a neighbourhood N of e in G such that σ is continuous in N x N. Then there exists a Borel function a(g) defined on G and taking values in T which is continuous in a neighbourhood of e in G such that $\sigma(g_1,g_2) = a(g_1)\, a(g_2)\, a(g_1 g_2)^{-1}$ for all $g_1, g_2 \in G$.

Proof: Since σ is a trivial multiplier it follows that there exists a Borel function a(g) on G such that $|a(g)| = 1$ and

$\sigma(g_1,g_2) = a(g_1)\, a(g_2)\, a(g_1g_2)^{-1}$. Consider the map $(g,\lambda) \to a(g)\lambda$ from G^σ into T. It is clear that this map is a Borel homomorphism and hence by Theorem 2.2 is a continuous homomorphism. By Lemma 3.1, the map $(g,\lambda) \to \lambda$ is continuous in a neighbourhood of the identity in G^σ. Hence the map $(g,\lambda) \to a(g)$ is continuous in a neighbourhood of the identity in G^σ. An application of Lemma 3.1 completes the proof.

We shall now prove a theorem connecting the local and global trivialities of a multiplier in a simply connected group.

<u>Theorem 3.2</u> Let G be a connected and simply connected locally compact second countable group. Suppose σ is a locally trivial multiplier on G x G. Then σ is globally trivial.

<u>Proof:</u> We may assume without loss of generality that there exists a neighbourhood N of e in G such that

$$\sigma(g_1, g_2) = 1 \qquad \text{if } g_1, g_2 \in N.$$

Consider the map θ from the cartesian product G x R of G and the real line onto G^σ defined by

$$\theta(g, r) = (g, e^{ir}).$$

Then θ is a continuous local homomorphism on the open set N x R. Since G is simply connected, G x R is also simply connected.

Hence there exists a continuous homomorphism θ_1 of G x R into G^σ such that

$$\theta(g, r) = \theta_1(g, r)$$

in an open subset of G x R. Let

$$\theta_1(g, 0) = \gamma(g).$$

Then $g \to \gamma(g)$ is a homomorphism from G into G^σ. In a neighbourhood of the identity, $\gamma(g)$ is of the form (g,1). If τ is defined by (3.1) we have

$$\tau(\gamma(g)) = g$$

in a neighbourhood of the identity. Since G is connected, it is generated by this neighbourhood. Hence $\tau(\gamma(g))=g$ for all g. Thus $\gamma(g)$ is of the form $(g,a(g))$ where $|a(g)|=1$. Since

$$\gamma(g_1g_2) = \gamma(g_1)\,\gamma(g_2) \quad \text{for all } g_1, g_2 \in G^\sigma,$$

we have

$$(g_1g_2,\ a(g_1g_2)) = (g_1g_2,\ a(g_1)a(g_2)\,\sigma(g_1,g_2)) \quad \text{for all } g_1, g_2 \in G^\sigma.$$

Hence

$$\sigma(g_1, g_2) = a(g_1g_2)\,a(g_1)^{-1}\,a(g_2)^{-1} \quad \text{for all } g_1, g_2 \in G^\sigma.$$

In other words σ is globally trivial.

4. MULTIPLIERS ON LIE GROUPS

Throughout this section we shall assume that G is a connected Lie group.

Theorem 4.1 (Bargmann [1])

Let G be a connected Lie group and σ a multiplier on G x G. Then there exists a multiplier σ' which is equivalent to σ and infinitely differentiable in a neighbourhood of the identity in G x G.

Proof: By Theorem 3.1 we may assume without loss of generality that there exists a compact neighbourhood N of e in G such that σ is continuous in N x N. Hence we can construct a Borel function $s(g_1,g_2)$ on G x G such that

$$\sigma(g_1,g_2) = \exp i\, s(g_1,g_2) \quad \text{for all } g_1,g_2\in G$$

where s is bounded in G x G and continuous in a neighbourhood of the form $N_1 \times N_1$, $e\in N_1$, $N_1 \subseteq N$. From the definition of a multiplier it follows that we can choose a neighbourhood N_2 of e in G such that $N_2 N_2 \subseteq N_1$ and

$$s(g_1,g_2) + s(g_1g_2,g_3) = s(g_1,g_2g_3) + s(g_2,g_3) \quad \text{for all } g_1,g_2,g_3\in N_2. \tag{4.1}$$

We shall denote by μ and ν the right and left invariant Haar measures of G. Let a be a C^∞ function such that

$$\left.\begin{aligned} a(g) &= 0 \qquad \text{if } g \notin N_2 \\ \int a(g)\, d\mu(g) &= 1 . \end{aligned}\right\} \qquad (4.2)$$

Let

$$\phi(g) = \int s(h, g)\, a(h)\, d\mu(h).$$

Consider

$$\tilde{s}(g_1,g_2) = s(g_1,g_2) - \phi(g_1) - \phi(g_2) + \phi(g_1 g_2).$$

It is clear that $\tilde{s}$ satisfies (4.1). By (4.2)

$$\tilde{s}(g_1,g_2) = \int_{N_2} [s(g_1,g_2) - s(h,g_1) - s(h,g_2) + s(h,g_1 g_2)] a(h)\, d\mu(h)$$

for all g_1,g_2.

If $g_1,g_2 \varepsilon N_2$, we have from (4.1), and the right invariance of μ

$$\begin{aligned} \tilde{s}(g_1,g_2) &= \int_{N_2} [s(hg_1,g_2) - s(h,g_2)]\, a(h)\, d\mu(h) \\ &= \int [s(hg_1,g_2) - s(h,g_2)]\, a(h)\, d\mu(h) \\ &= \int s(h,g_2)\, [a(h\, g_1^{-1}) - a(h)]\, d\mu(h) \end{aligned}$$

for all $g_1,g_2 \varepsilon N_2$. (4.3)

Let $N_3 \subseteq N_2$ be a neighbourhood of e such that $N_3\, N_3 \subseteq N_2$.

Select a C^∞ function b such that

$$\left.\begin{aligned} b(g) &= 0 \quad \text{if } g \notin N_3 \\ \int b(g)\, d\nu(g) &= 1. \end{aligned}\right\} \qquad (4.4)$$

Let

$$\psi(g) = \int \tilde{s}(g, k)\, b(k)\, d\nu(k).$$

Define

$$\tilde{\tilde{s}}(g_1, g_2) = \tilde{s}(g_1, g_2) - \psi(g_1) - \psi(g_2) + \psi(g_1\, g_2).$$

Then $\tilde{\tilde{s}}$ satisfies (4.1) for all $g_1, g_2, g_3 \varepsilon N_2$. Further if $g_1, g_2 \varepsilon N_2$, we have from the left invariance of ν,

$$\begin{aligned} \tilde{\tilde{s}}(g_1, g_2) &= \int_{N_2} \left[\tilde{s}(g_1, g_2) - \tilde{s}(g_1, k) - \tilde{s}(g_2, k) + \tilde{s}(g_1 g_2, k)\right] b(k)\, d\nu(k) \\ &= \int_{N_2} \left[\tilde{s}(g_1, g_2 k) - \tilde{s}(g_1, k)\right]\, b(k)\, d\nu(k) \\ &= \int \left[\tilde{s}(g_1, g_2\, k) - \tilde{s}(g_1, k)\right] b(k)\, d\nu(k) \\ &= \int \tilde{s}(g_1, k) \left[b(g_2^{-1}\, k) - b(k)\right]\, d\nu(k). \end{aligned}$$

Because of (4.4) it follows that the integrand in the last integral of the above equation vanishes for all $k \notin g_2 N_3 \cup N_3$. Since $N_3\ N_3 \subseteq N_2$ it follows that

$$\tilde{\tilde{s}}(g_1, g_2) = \int_{N_2} \tilde{s}(g_1, k) \left[b(g_2^{-1}\, k) - b(k)\right] d\nu(k)$$

for all $g_1, g_2 \varepsilon N_3$.

Hence by (4.3)

$$\tilde{\tilde{s}}(g_1,g_2) = \iint s(h, k) \left[a(h\, g_1^{-1}) - a(h)\right] \left[b(g_2^{-1}k) - b(k)\right] d\mu(h)\, d\nu(k)$$

for all $g_1,g_2 \in N_3$.

Since a and b are C^∞ functions vanishing outside a compact set it is clear that $\tilde{\tilde{s}}$ is infinitely differentiable in $N_3 \times N_3$. From the construction of $\tilde{s}$ and $\tilde{\tilde{s}}$ it is evident that $\sigma = e^{is} \sim e^{i\tilde{s}} \sim e^{i\tilde{\tilde{s}}}$. This completes the proof.

Suppose G is a connected Lie group and σ is a multiplier on G x G which is infinitely differentiable in a neighbourhood of the identity. We shall now show that the group G^σ itself admits a Lie structure.

Consider a coordinate neighbourhood N of the identity in G where for any $g \in N$, the coordinates are given by $x_1(g),\ldots,x_n(g)$. Let $N_1 \subset N$ be a open neighbourhood of the identity such that $N_1 N_1^{-1} \subset N$ and the multiplication operation

$$(g_1, g_2) \to g_1\, g_2^{-1}, \qquad g_1,g_2 \in N_1$$

is given by infinitely differentiable functions

$$x_i(g_1 g_2^{-1}) = f_i(x_1(g_1),\ldots,x_n(g_1);\ x_1(g_2),\ldots,x_n(g_2)), \quad i=1,2,\ldots,n.$$

We may further assume that N is so small that σ is infinitely differentiable in N x N and that $\sigma(g_1,g_2) = \exp i\, s(g_1,g_2)$ for $g_1,g_2 \in N$ where s is infinitely differentiable. Let

$$M_\varepsilon = \{e^{ir}: |r| < \varepsilon\} \qquad \text{for } \varepsilon > 0.$$

Let N be so small that

$$|s(g_1, g_2)| < \varepsilon/4 \qquad \text{for all } g_1, g_2 \varepsilon N.$$

Consider the coordinate system given by the map

$$\tau: (g, e^{ir}) \to x_1(g), \ldots x_n(g), r$$

from $N \times M_\varepsilon$ into R^{n+1}. By the remark after Lemma 3.3 it follows that τ is a homeomorphism of a neighbourhood of the identity in G^σ into R^{n+1} if N is sufficiently small and ε is also sufficiently small. Let now (g_1, e^{ir_1}), (g_2, e^{ir_2}) be in $N_1 \times M_{\varepsilon/4}$. Then

$$(g_1, e^{ir_1})(g_2, e^{ir_2})^{-1} = (g_1 g_2^{-1}, e^{i[(r_1-r_2)+s(g_1,g_2^{-1})-s(g_2,g_2^{-1})]})$$

Thus

$$\tau((g_1, e^{ir_1})o(g_2, e^{ir_2})^{-1}) =$$

$$(x_1(g_1 g_2^{-1})\ldots x_n(g_1 g_2^{-1}), \ r_1-r_2 + s(g_1,g_2^{-1}) - s(g_2,g_2^{-1}))$$

Since all the entries on the right hand side of the above equation are infinitely differentiable in the variables

$$(x_1(g_1), \ldots, x_n(g_1), r_1, \ x_1(g_2), \ldots, x_n(g_2), r_2)$$

it follows that the multiplication operation in G^σ is infinitely differentiable in a neighbourhood of the identity. Hence by a well known theorem G^σ itself admits a Lie structure.

The subgroup T^σ defined in §3 is compact and connected. By

Lemma 3.1, G^σ/T^σ is also connected. Hence G^σ becomes a connected Lie group.

In particular, if G is an analytic Lie group G^σ also admits an analytic structure. The mapping τ: $(g,\lambda) \to g$ is a continuous and therefore analytic homomorphism from G^σ onto G. By a theorem of Chevalley it follows that there exists an analytic map γ of an open set N containing e in G into G^σ such that

$$\tau (\gamma(g)) = g \qquad \text{for all } g \epsilon N.$$

i.e., $\gamma(g)$ is of the form

$$\gamma(g) = (g, a(g)) \qquad \text{for all } g \epsilon N,$$

where $|a(g)| = 1$. We have

$$\gamma(g_1)\, \gamma(g_2)\, \gamma(g_1 g_2)^{-1} = (e,\, a(g_1)\, a(g_2)\, a(g_1 g_2)^{-1}\, \sigma(g_1,g_2))$$
$$\text{for all } g_1,g_2 \epsilon N.$$

If N_1 is a neighbourhood of e such that $N_1.N_1 \subset N$, it follows that $\sigma(g_1,g_2)\, a(g_1)\, a(g_2)\, a(g_1 g_2)^{-1}$ is analytic in $N_1 \times N_1$. Thus we have the following corollary.

<u>Corollary 1</u> (Varadarajan [13]). If G is an analytic Lie group and σ is a multiplier on G x G, then there exists a multiplier σ' which is equivalent to σ and analytic in a neighbourhood of the identity in G x G.

We conclude this section with a lemma which we shall have occasion to use later.

<u>Lemma 4.1</u> Let G be a connected (analytic) Lie group and σ be a trivial multiplier on G x G. Suppose σ is infinitely differentiable (analytic) in a neighbourhood of the identity in G x G. Then there exists a Borel function a(g) on G such that $|a(g)| = 1$, a(g) is infinitely differentiable(analytic)in a neighbourhood of e and $\sigma(g_1,g_2) = a(g_1)a(g_2)a(g_1 g_2)^{-1}$

for all $g_1, g_2 \varepsilon G$.

Proof: We shall prove the lemma only in the infinitely differentiable case. The analytic case is proved in the same manner. By the remarks made before the statement of Corollary 1 to Theorem 4.1 it follows that G^σ can be considered as a connected Lie group. Since σ is trivial there is a Borel function $a(g)$ such that $\sigma(g_1, g_2) = a(g_1)\, a(g_2)\, a(g_1 g_2)^{-1}$ for all $g_1, g_2 \varepsilon G$. Consider the map $(g, \lambda) \to a(g)\lambda$ from G^σ into T. This is a Borel homomorphism from G^σ into T. Hence by Theorem 2.2 it is continuous and therefore infinitely differentiable. In the proof of Lemma 3.3 if we take f_o to be an infinitely differentiable function vanishing outside a suitably chosen compact set it follows that the map $(g, \lambda) \to \lambda$ is infinitely differentiable in a neighbourhood of the identity in G^σ. Thus the map $(g, \lambda) \to a(g)$ is infinitely differentiable in a neighbourhood of the identity. Since the quotient Lie structure of G is the same as its original Lie structure the map $g \to a(g)$ is infinitely differentiable in a neighbourhood of e in G. This completes the proof.

5. MULTIPLIERS ON SOME SPECIAL GROUPS

In this section we shall study some of the properties of multipliers in the case of compact groups, abelian groups, semisimple Lie groups and semidirect products of vector spaces and groups of endomorphisms on them.

a) The case of a compact group

Theorem 5.1 (Bargmann [1])
Let G be a compact second countable group and σ a multiplier on G x G. Then there exists an integer k such that σ^k is trivial. Further σ is locally trivial.

Proof: By Theorem 3.1 we may assume that σ is continuous in a neighbourhood of the identity. Since G is compact it has a totally finite right (and left) invariant measure μ. Hence the right invariant Haar measure on the locally compact group G^σ is totally finite. Hence by Corollary 1 to Theorem 2.1, G^σ is compact in the Weil topology. Consider the unitary operators U_g and $U_{(g,\lambda)}$ defined by (3.2). By Lemma 3.2, $(g,\lambda) \to U_{(g,\lambda)}$ is a weakly continuous unitary representation of G^σ. Hence by Peter-Weyl theorem it follows that there exists a non trivial finite dimensional subspace $\mathcal{V} \subseteq \mathcal{L}_2(\mu)$ which is invariant under all $U_{(g,\lambda)}$ and hence all U_g, $g \epsilon G$. Let $\dim \mathcal{V} = k$. Let d(g) be the determinant of U_g in $\mathcal{V}$. Since $U_{g_1} U_{g_2} = \sigma(g_1,g_2) U_{g_1 g_2}$ it follows that $d(g_1)\, d(g_2) = \sigma(g_1,g_2)^k\, d(g_1 g_2)$. In other words σ^k is trivial. Since σ^k is continuous we may assume by Lemma 3.4 that d(g) is continuous in a neighbourhood of the identity in G. Hence there exists a Borel function a(g) which is continuous in a neighbourhood of the identity in G and satisfies the equation $a(g)^k = d(g)$. Thus $\sigma(g_1,g_2) = a(g_1)\, a(g_2)\, a(g_1 g_2)^{-1}$ for all g_1, g_2 in a neighbourhood of the identity. This completes the proof.

Corollary 1 If G is a compact, connected and simply connected second countable group then every multiplier on $G \times G$ is trivial.

Proof: This is an immediate consequence of Theorems 5.1 and 3.2.

b) Abelian groups

Definition 5.1 Let G be a locally compact group and σ be a multiplier on $G \times G$. σ is said to be symmetric if $\sigma(g_1, g_2) = \sigma(g_2, g_1)$ for all $g_1, g_2 \varepsilon G$.

Theorem 5.2 Let G be a locally compact second countable abelian group and σ be a symmetric multiplier on $G \times G$. Then there exists an integer k such that σ^k is trivial. Further σ is locally trivial.

Proof: The symmetry of σ implies that G^σ is abelian. Choose and fix a point $(e, \lambda_o) \varepsilon G^\sigma$ where $\lambda_o \neq 1$ and $|\lambda_o| = 1$. There exists a character χ on G^σ such that $\chi(e, \lambda_o) \neq 1$. Since

$$(g, \lambda) = (g, 1) \circ (e, \lambda) \quad \text{for all } (g, \lambda) \varepsilon G^\sigma,$$

it follows that $\chi(g, \lambda)$ is of the form

$$\chi(g, \lambda) = a(g)\, b(\lambda)$$

where $a(g) = \chi(g, 1)$, $b(\lambda) = \chi(e, \lambda)$. Since $(e, \lambda_1) \circ (e, \lambda_2) = (e, \lambda_1 \lambda_2)$, we have $b(\lambda_1)\, b(\lambda_2) = b(\lambda_1 \lambda_2)$ for all $\lambda_1, \lambda_2 \varepsilon T$. Since $b(\lambda)$ is a Borel function on T and $b(\lambda_o) \neq 1$, it follows that there exists an integer $k \neq 0$ such that $b(\lambda) = \lambda^k$ for all $\lambda \in T$. Thus

$$\chi(g, \lambda) = a(g)\, \lambda^k, \qquad \text{for all } (g,\lambda) \in G^\sigma.$$

Since χ is a character on G^σ, we have

$$a(g_1)\, a(g_2)\, \lambda_1^k\, \lambda_2^k = a(g_1 g_2)\, \sigma(g_1, g_2)^k\, \lambda_1^k\, \lambda_2^k$$

$$\text{for all } (g_1, \lambda_1), (g_2, \lambda_2) \in G^\sigma.$$

This proves the first part of the theorem. The second part is proved exactly as in the case of Theorem 5.1.

<u>Theorem 5.3</u> Let R be the real line and σ be a multiplier on R x R. Then σ is trivial.

<u>Proof</u>: Since R is a connected Lie group it follows from Theorem 4.1 that we may assume σ to be infinitely differentiable in a neighbourhood of e and G^σ to be a connected Lie group. For all sufficiently small $x \in R$, the point $(x,1) \in R^\sigma$ lies on a continuous one parameter subgroup (x_t, λ_t), $t \in R$. Choose and fix such an $x \neq 0$. Suppose $x_{t_o} = x$. The equation

$$(x_t, \lambda_t) o (x_s, \lambda_s) = (x_t + x_s, \lambda_t \lambda_s\, \sigma(x_t, x_s))$$

$$= (x_{t+s}, \lambda_{t+s}) \quad \text{for all } t,s \in R.$$

implies that

$$x_t + x_s = x_{t+s};\ \sigma(x_t, x_s) = \lambda_{t+s}\, \lambda_t^{-1}\, \lambda_s^{-1} \quad \text{for all } t,s \in R$$

From continuity it follows that $x_t = t_o^{-1} tx$. Thus

$$\sigma(t_o^{-1}\, tx,\ t_o^{-1}\, sx) = \lambda_{t+s}\, \lambda_t^{-1}\, \lambda_s^{-1} \qquad \text{for all } t,s\in R.$$

Since every point of the real line is of the form t_o^{-1} tx for some t, this completes the proof.

Theorem 5.4 (Bargmann [1])

Let V be a real finite dimensional vector space and σ be a multiplier on V x V. Then there exists a real skew symmetric bilinear function B(x,y) on V x V such that the multiplier defined by the function exp i B(x,y) is equivalent to σ.

Proof: It is trivial to verify that the function exp i B(x,y) is a multiplier for every real bilinear function on V x V.

Because of Theorem 4.1 we may assume that σ is infinitely differentiable in a neighbourhood of the origin in V x V and V^σ is a connected Lie group. For any $x\in V$ consider the subgroup of all points tx, $t\in R$. By Theorem 5.3, it follows that there exists a function $\lambda_t(x)$ such that

$$\left.\begin{aligned} \sigma(tx,\ sx) &= \lambda_{t+s}(x)\, \lambda_t(x)^{-1}\, \lambda_s(x)^{-1} \\ |\lambda_t(x)| &= 1 \end{aligned}\right\} \quad \text{for all } t,s\in R,\ x\in V,\ x\neq 0 \tag{5.1}$$

In other words $\{(tx, \lambda_t(x)),\ t\in R\}$ is a one parameter subgroup of V^σ. Hence for any $y\in V$, $\{(y,1)o(tx,\lambda_t(x)o(y,1)^{-1},\ t\in R\}$ is also a one parameter subgroup of V^σ. But an easy calculation shows that

$$(y, 1)o(tx, \lambda_t(x))o(y, 1)^{-1} = (tx, \mu_t)$$

where

$$\mu_t = \lambda_t(x)\, \sigma(y, tx)\, \sigma(y+tx, -y)\, \sigma(y, -y)^{-1}.$$

Since $\{(tx, \mu_t), t\varepsilon R\}$ is a one parameter subgroup, we obtain

$$\sigma(tx, sx) = \mu_{t+s}\, \mu_t^{-1}\, \mu_s^{-1}, \quad t,s\varepsilon R. \tag{5.2}$$

From (5.1) and (5.2) we conclude that $\mu_t\, \lambda_t(x)^{-1}$ is a one parameter group in t. Since $\mu_t\, \lambda_t(x)^{-1}$ is a Borel function in t, it follows that there exists a function f(x,y) such that

$$\mu_t\, \lambda_t(x)^{-1} = \sigma(y,tx)\, \sigma(y + tx, -y)\, \sigma(y, -y)^{-1} = \exp it\, f(x,y) \tag{5.3}$$

for all $t\varepsilon R$.

Since σ is a locally C^∞ function there exists a neighbourhood N of the origin in V such that

$$\left.\frac{d\, \sigma(y,tx)}{dt}\right|_{t=0} + \sigma(y, -y)^{-1} \left.\frac{d\, \sigma(y+tx,-y)}{dt}\right|_{t=0} = if(x,y) \tag{5.4}$$

for all $y\varepsilon N$, $x\varepsilon V$.

It follows in particular that f(x,y) is linear in x. Putting t=1 in (5.3) we obtain

$$\sigma(y, x)\, \sigma(y + x, -y) = \sigma(y, -y) \exp if(x,y).$$

But

$$\sigma(x,y)\, \sigma(x + y, -y) = \sigma(y, -y).$$

From the above equations

$$\sigma(x, y)\ \sigma(y, x)^{-1} = \exp i\ f(x, y) \qquad \text{for all } x,y\varepsilon V.$$

Thus $f(x,y) = - f(y,x)$. In other words f is a skew symmetric real bilinear functional on V x V. Since V is simply connected it follows from Theorems 3.2 and 5.2 that the symmetric multiplier $\sigma(x,y)\ \sigma(y,x)$ is trivial. Hence

$$\begin{aligned} \sigma^2(x, y) &= \sigma(x,y)\ \sigma(y,x)^{-1}\ \sigma(y,x)\ \sigma(x,y) \\ &\sim \sigma(x,y)\ \sigma(y,x)^{-1} = \exp - i\ f(x,y). \end{aligned}$$

Hence $\sigma(x,y) \exp \frac{i}{2} f(x,y)$ is locally and hence globally trivial. If we choose $B(x,y) = - \frac{1}{2} f(x,y)$ this completes the proof of the theorem.

c) Semi simple Lie groups

We shall first prove a special case of a result due to J. H. C. Whitehead (cf Theorem[14], page 96, [6]. See also Bargmann [1]).

Theorem 5.5 (J. H. C. Whitehead)

Let G be a connected semi simple Lie group and σ be a multiplier on G x G. Then σ is locally trivial.

Proof: By Theorem 4.1 we may assume that σ is infinitely differentiable in a neighbourhood of the identity in G x G and G^σ is a connected Lie group. Let $\mathfrak{g}$ and $\mathfrak{g}^\sigma$ be the Lie algebras of G and G^σ respectively. As before we shall denote by τ the homomorphism $(g,\lambda) \to g$ from G^σ onto G. The kernel of τ is the subgroup $T^\sigma = \{(e,\lambda),\ \lambda\varepsilon T\}$. T^σ is in the centre of G^σ. Hence $d\tau\colon \mathfrak{g}^\sigma \to \mathfrak{g}$

is a homomorphism from $\mathfrak{g}^{\sigma}$ onto $\mathfrak{g}$ whose kernel is the Lie subalgebra $\mathfrak{T}^{\sigma}$ of the subgroup T^{σ}. Since T^{σ} is a one dimensional Lie group, $\mathfrak{T}^{\sigma}$ is one dimensional. Let it be generated by $X_0 \neq 0$. Let $\mathfrak{L}$ be any complementary subspace for $\mathfrak{T}^{\sigma}$ in $\mathfrak{g}^{\sigma}$ so that the vector space $\mathfrak{g}^{\sigma}$ can be written as the direct sum $\mathfrak{T}^{\sigma} \dotplus \mathfrak{L}$. The map $d\tau: \mathfrak{L} \rightarrow \mathfrak{g}$ is a linear isomorphism onto $\mathfrak{g}$. Let $\eta: \mathfrak{g} \rightarrow \mathfrak{L}$ be the inverse of the map $d\tau$ restricted to $\mathfrak{L}$. Let $[\cdot,\cdot]$ and $[\cdot,\cdot]^{\sigma}$ denote the Lie brackets in $\mathfrak{g}$ and $\mathfrak{g}^{\sigma}$ respectively.

Let $\tilde{\mathfrak{g}} = \mathfrak{g} \times R$ be the direct product of the vector spaces $\mathfrak{g}$ and the real line R. Consider the mapping

$$\gamma : \tilde{\mathfrak{g}} \longrightarrow \mathfrak{g}^{\sigma}$$

defined by

$$\gamma(X, c) = c\, X_0 + \eta(X), \qquad X \in \mathfrak{g}, \; c \in R.$$

γ is then a 1 - 1 linear isomorphism. We shall convert $\tilde{\mathfrak{g}}$ into a Lie algebra as follows: define the Lie bracket $[\cdot,\cdot]^{\sim}$ in $\tilde{\mathfrak{g}}$ by

$$[(X_1, c_1), (X_2, c_2)]^{\sim} = \gamma^{-1}\{[\gamma(X_1, c_1), \gamma(X_2, c_2)]^{\sigma}\} .$$

It is clear that $\tilde{\mathfrak{g}}$ is now a Lie algebra isomorphic to $\mathfrak{g}^{\sigma}$. Since X_0 is in the centre of $\mathfrak{g}^{\sigma}$ it follows that

$$[\gamma(X_1, c_1), \gamma(X_2, c_2)]^{\sigma} = [\eta(X_1), \eta(X_2)]^{\sigma} ,.$$

Since $d\tau$ is a homomorphism from $\mathfrak{g}^{\sigma}$ onto $\mathfrak{g}$,

$$d\tau\,[\eta(X_1),\ \eta(X_2)]^{\sigma} = [d\tau.\eta(X_1),\ d\tau.\eta(X_2)] = [X_1,\ X_2]\,.$$

But

$$d\tau.\eta\,[X_1,\ X_2] = [X_1,\ X_2]\,.$$

Hence

$$[\eta(X_1),\ \eta(X_2)]^{\sigma} = \eta[X_1,\ X_2] + B(X_1,\ X_2)\,X_o$$

$$\text{for all } X_1, X_2 \in \mathfrak{g}\,,$$

where $B(X_1,X_2)$ is a real scalar. Thus by our definition of the Lie algebra $\tilde{\mathfrak{g}}$, we have

$$[(X_1,\ c_1),\ (X_2,\ c_2)]^{\sim} = ([X_1,\ X_2],\ B(X_1,X_2)\ ),$$

$$X_1,\ X_2 \in \mathfrak{g}$$
$$c_1,\ c_2 \in R.$$

It is clear that B cannot be arbitrary since $[.,.]^{\sim}$ is a Lie bracket. The skew symmetry and bilinearity of $[.,.]^{\sim}$ implies that $B(X_1,X_2)$ is skew symmetric and bilinear in X_1 and X_2. The Jacobian identity for $[.,.]^{\sim}$ implies that

$$B([X,\ Y],\ Z) + B([Y,\ Z],\ X) + B([Z,\ X],\ Y) = 0$$

$$\text{for all } X,Y,Z \in \mathfrak{g} \qquad (5.5)$$

Till now our discussion was quite general and we have not made use of the fact that $\mathfrak{g}$ is semi simple. We shall now do this.

Let $C(X,Y)$ be the Cartan-Killing form of $\mathfrak{g} \times \mathfrak{g}$. Since $\mathfrak{g}$ is semi simple, $C(.,.)$ is non singular. Hence for every $Y \varepsilon \mathfrak{g}$ there exists a $\beta(Y) \varepsilon \mathfrak{g}$ such that

$$B(X, Y) = C(X, \beta(Y)) \quad \text{for all } X \varepsilon \mathfrak{g} . \tag{5.6}$$

It is well known that $C(.,.)$ satisfies the identity

$$C([X, Y], Z) = C(X, [Y, Z]).$$

Hence from (5.5) and (5.6) we have

$$C([X, Y], \beta(Z)) + C([Y, Z], \beta(X)) + C([Z, X], \beta(Y)) = 0. \tag{5.7}$$

Skew symmetry of $B(X, Y)$ imples that

$$C(X, \beta(Y)) = - C(\beta(X), Y). \tag{5.8}$$

From (5.7) and (5.8) we obtain

$$C(Y, \beta([X, Z])) = C(Y, [\beta(X), Z]) + C(Y, [X, \beta(Z)])$$

$$\text{for all } X,Y,Z \varepsilon \mathfrak{g} .$$

Since C is non singular,

$$\beta([X, Z]) = [\beta(X), Z] + [X, \beta(Z)] .$$

i.e., the mapping $X \rightarrow \beta(X)$ is a derivation of $\mathfrak{g}$. Since $\mathfrak{g}$ is semi simple every derivation is inner. Hence there exists a $Y_0 \varepsilon \mathfrak{g}$ such that

$$\beta(Y) = [Y_o, Y] \qquad \text{for all } Y \in \mathfrak{g} .$$

Hence from (5.6) we get

$$B(X, Y) = C(X, [Y_o, Y]) = - C(Y_o, [X, Y])$$

$$= \Lambda([X, Y]), \qquad \text{say},$$

where Λ is a linear functional on $\mathfrak{g}$.

$\tilde{\mathfrak{g}}$ is a Lie algebra isomorphic to $\mathfrak{g}^{\sigma}$. In $\tilde{\mathfrak{g}}$ the Lie bracket is given by

$$[(X_1, c_1), (X_2, c_2)]^{\sim} = ([X_1, X_2], \Lambda([X_1, X_2])).$$

Consider the Lie algebra $\bar{\mathfrak{g}}$ which is the direct sum of $\mathfrak{g}$ and R. Here the Lie bracket $[\cdot,\cdot]^{-}$ is given by

$$[(X_1, c_1), (X_2, c_2)]^{-} = ([X_1, X_2], 0).$$

Consider the map π: $\tilde{\mathfrak{g}} \to \mathfrak{g}$ defined by

$$\pi(X, c) = (X, c - \Lambda(X)).$$

π is obviously a linear isomorphism.

$$\pi [(X_1, c_1), (X_2, c_2)]^{\sim} = \pi ([X_1, X_2], \Lambda([X_1, X_2]))$$

$$= ([X_1, X_2], 0)$$

$$= [(X_1, c_1 - \Lambda(X_1)), (X_2, c_2 - \Lambda(X_2))]^{-}$$

$$= [\pi(X_1, c_1), \pi(X_2, c_2)]^{-}.$$

Thus π is a Lie algebra isomorphism between $\tilde{\mathfrak{g}}$ and $\bar{\mathfrak{g}}$. Further π takes the projection map $(X, c) \to X$ in $\tilde{\mathfrak{g}}$ to the projection map $(X, c) \to X$ in $\bar{\mathfrak{g}}$. Hence there is a local isomorphism ϕ between the Lie groups G^σ and the direct product $G \times T$ (of G and the unit circle in the complex plane) such that

$$\phi(g, \lambda) = (g, a(g,\lambda))$$

for all (g, λ) in a neighbourhood of the identity in G^σ. In particular we obtain

$$a(g_1, \lambda_1)\, a(g_2, \lambda_2) = a(g_1 g_2, \lambda_1 \lambda_2\, \sigma(g_1, g_2)) \tag{5.9}$$

for all (g_1, λ_1), (g_2, λ_2) in a neighbourhood N of $(e,1)$ in G^σ. Since $(g,\lambda) = (g,1)\circ(e,\lambda)$ in G^σ,

$$a(g,\lambda) = a(g, 1)\, a(e, \lambda) \qquad \text{for all } (g, \lambda) \in N. \tag{5.10}$$

Further

$$a(e, \lambda_1)\, a(e, \lambda_2) = a(e, \lambda_1 \lambda_2) \qquad \text{for all } (e,\lambda_1), (e,\lambda_2) \in N.$$

Hence $a(e,\lambda) = \lambda^n$ for all $(e,\lambda) \in N$ for some integer n. Equations (5.9) and (5.10) now imply

$$\sigma(g_1, g_2)^n = a(g_1)\, a(g_2)\, a(g_1 g_2)^{-1} \qquad \text{for all } g_1, g_2 \in V$$

where $a(g) = a(g,1)$ and V is some neighbourhood of e in G. Since $a(g)$ is locally continuous and so is σ, we conclude that σ is locally trivial. This completes the proof.

From Theorems 5.5 and 3.2 we now obtain

Corollary 1 If G is a connected and simply connected semisimple Lie group, then every multiplier on G x G is trivial.

d) The case of semi direct products

Let G be a locally compact second countable group, N a closed normal subgroup and H a closed subgroup of G such that $N \cap H = \{e\}$. Then every element $g \in G$ can be written as nh in a unique manner where $n \in N$ and $h \in H$. The multiplication operation between two elements n_1h_1 and n_2h_2 can be written as $n_1h_1\, n_2h_2 = n_1(h_1\, n_2\, h_1^{-1})h_1h_2$ for all $n_1, n_2 \in N$ and $h_1, h_2 \in H$. G is called a semidirect product of N and H. The mapping $n \to h\, n\, h^{-1}$ is an automorphism of N for every $h \in H$. We shall write $h(n)$ for $h\, n\, h^{-1}$.

It is clear that direct products of two groups are examples of semidirect products. If N is a locally compact second countable group and H is a locally compact group of automorphisms of N in the compact open topology then the product topological space N x H can be endowed with the group operation $(n_1, h_1)o(n_2, h_2) = (n_1\, h_1(n_2),\, h_1h_2)$ for all $n_1, n_2 \in N$ and $h_1, h_2 \in H$. This is an example of a locally compact group which is a semidirect product.

We shall now investigate the problem of expressing the multipliers of the semidirect product of N and H in terms of the multipliers of N and H. The main result is due to Mackey (cf Theorem 9.4, [8])

<u>Theorem 5.6</u> (Mackey [8])

Let G be a locally compact second countable group which is the semidirect product of a closed normal subgroup N and a closed subgroup H such that $N \cap H = \{e\}$. Let ω be a multiplier on G x G. Then there exists an equivalent multiplier ω_1 of the form

$$\omega_1(n_1h_1, n_2h_2) = \sigma(n_1,h_1(n_2)) \ \delta(h_1,h_2) \ \psi(n_2,h_1)$$

$$\text{for all } n_1,n_2 \varepsilon N, \ h_1h_2 \varepsilon H \qquad (5.11)$$

where σ is a multiplier on N x N, δ is a multiplier on H x H and ψ is a Borel function defined on N x H and taking values in T and σ, δ and ψ satisfy the following conditions:

(i) $\sigma(h(n_1),h(n_2) = \sigma(n_1,n_2) \ \psi(n_1n_2,h)\psi(n_1,h)^{-1}\psi(n_2,h)^{-1}$

for all $h\varepsilon H$, $n_1,n_2\varepsilon N$,

(ii) $\psi(n, h_1h_2) = \psi(h_2(n),h_1) \ \psi(n,h_2)$ for all $n\varepsilon N$, $h_1,h_2 \ \varepsilon \ H$.

Conversely if σ, δ and ψ are functions satisfying the conditions described above, then the function ω_1 defined by (5.11) is a multiplier on G x G.

<u>Proof:</u> The converse part of the theorem is proved by direct verification and therefore left to the reader.

Throughout the proof we shall denote by n and h with suffixes the elements of N and H respectively. We deduce from the definition of the multiplier that

$$\omega(n_1h_1,\ n_2h_2) = \omega(n_1,\ h_1(n_2)) \frac{\omega(n_1h_1(n_2),\ h_1h_2)\ \omega(h_1,\ n_2h_2)}{\omega(n_1,\ h_1)\ \omega(h_1(n_2),\ h_1h_2)} .$$

If we denote by $\alpha(nh)$ the function $\omega(n,h)$, we can rewrite the above equation as

$$\omega(n_1h_1,\ n_2h_2) = \omega(n_1,\ h_1(n_2)) \frac{\alpha(n_1h_1\ n_2h_2)}{\alpha(n_1h_1)\ \alpha(n_2h_2)} \quad \omega(h_1,h_2) \frac{\omega(h_1,n_2)}{\omega(h_1(n_2),h_1)}$$

$$\sim \omega(n_1,\ h_1(n_2))\ \omega(h_1,h_2)\ \psi(n_2,h_1)$$

where $\psi(n_2,h_1) = \omega(h_1,n_2)\ \omega(h_1(n_2),\ h_1)^{-1}$. We shall denote by σ and δ the restrictions of ω to N x N and H x H respectively. Then σ and δ are obviously multipliers. Thus ω is equivalent to a multiplier ω_1 where ω_1 is defined by

$$\omega_1(n_1h_1,\ n_2h_2) = \sigma(n_1,\ h_1(n_2))\ \delta(h_1,h_2)\ \psi(n_2,h_1). \tag{5.12}$$

Putting $h_1=n_1=e$; $h_1=h_2=e$; and $n_1=n_2=e$ successively in (5.12) we get

$$\psi(n_2,\ h_1) = \omega_1(h_1,\ n_2); \quad \sigma(n_1,\ n_2) = \omega_1(n_1,n_2); \quad \delta(h_1,h_2) = \omega_1(h_1,h_2) \tag{5.13}$$

Conditions (i) and (ii) can now be deduced directly from (5.13) and the fact that ω_1 is a multiplier.

Corollary 1 If G is the direct product of two locally compact groups N and H and ω is a multiplier on G x G, then there exists an equivalent multiplier ω_1 given by

$$\omega_1(n_1h_1, n_2h_2) = \sigma(n_1,n_2)\ \delta(h_1,h_2)\ \psi(n_2,h_1)$$

where σ and δ are multipliers on N and H respectively and ψ satisfies the equations

$$\psi(n_1n_2, h) = \psi(n_1,h)\ \psi(n_2,h) \quad \text{for all } n_1,n_2\varepsilon N,\ h\varepsilon H,$$

$$\psi(n, h_1h_2) = \psi(n,h_1)\ \psi(n,h_2) \quad \text{for all } h_1,h_2\varepsilon H,\ n\varepsilon N.$$

Remark 1 Since a vector space is a direct sum of copies of the real line it is clear from the above corollary that Theorem 5.4 follows immediately from Theorem 5.3.

Remark 2 If G is the direct sum of a vector space V and a compact group H, then every multiplier on G is equivalent to a multiplier ω of the form

$$\omega((n_1,h_1), (n_2,h_2)) = \exp iB(n_1,n_2)\ \ \delta(h_1,h_2)$$

$$\text{for all } n_1,n_2\varepsilon V,\ h_1,h_2\varepsilon K.$$

where B is a real skew symmetric bilinear form on V and δ is a multiplier on H.

As an application of Theorem 5.6 we shall illustrate two more examples which are of great interest in quantum mechanics. Let V be a real finite dimensional vector space and H a Lie group of endomorphisms of V. Let G be the semidirect product of V and H. Any element of G can be expressed as nh, $n \in V$, $h \in H$ and the "product" $n_1h_1\, n_2h_2$ is given by $(n_1+h_1(n_2))\, h_1h_2$.

Before proceeding to the statement of the main result we shall prove a lemma concerning a functional equation which arises in Mackey's discussion on multipliers ([10], page 286).

<u>Lemma 5.1</u> Let G be either a compact group or a connected semisimple Lie group of endomorphisms on a real finite dimensional vector space V. Let $\underset{\sim}{f}: G \to V$ be a Borel map which is analytic in a neighbourhood of the identity. Suppose

$$\underset{\sim}{f}(g_1\, g_2) = g_1\, \underset{\sim}{f}(g_2) + \underset{\sim}{f}(g_1) \quad \text{for all } g_1, g_2 \in G. \tag{5.14}$$

Then there exists a vector $\underset{\sim}{\alpha} \in V$ such that

$$\underset{\sim}{f}(g) = g\, \underset{\sim}{\alpha} - \underset{\sim}{\alpha} \quad \text{for all } g \in G.$$

<u>Proof:</u> First we shall consider the case when G is compact. Keep g_1 fixed and integrate both sides of (5.14) with respect to the normalised Haar measure. Then

$$\int \underset{\sim}{f}(g_2)\, dg_2 = \int \underset{\sim}{f}(g_1g_2)\, dg_2 = g_1 \int \underset{\sim}{f}(g_2)\, dg_2 + \underset{\sim}{f}(g_1).$$

If we write $\underset{\sim}{\alpha} = -\int \underset{\sim}{f}(g)dg$ it follows that

$$\underset{\sim}{f}(g) = g\,\underset{\sim}{\alpha} - \underset{\sim}{\alpha} \; .$$

This completes the proof when G is compact.

Let now G be connected and semi simple. Let $\mathfrak{g}$ be the Lie algebra of G and let the infinitesimal form of the representation $g \to g$ in V be denoted by π. We have for every $X \in \mathfrak{g}$,

$$\underset{\sim}{f}(g \operatorname{expt} X) = g\,\underset{\sim}{f}(\operatorname{expt} X) + \underset{\sim}{f}(g) \qquad \text{for all } t \tag{5.15}$$

Suppose $\underset{\sim}{f}$ is analytic in an exponential coordinate neighbourhood N of the identity. Differentiating (5.15) with respect to t at $t=0$, we obtain

$$(X\,\underset{\sim}{f})\,(g) = g\,(X\,\underset{\sim}{f})\,(e) \qquad \text{for all } g \in N,\ X \in \mathfrak{g} \tag{5.16}$$

Here X is considered as a left invariant vector field in G and is applied to every component of $\underset{\sim}{f}$. If we put $g = \operatorname{expt} Y$ in (5.16) and differentiate at $t=0$, we have

$$(Y\,X\,\underset{\sim}{f})\,(e) = \pi(Y)\,(X\,\underset{\sim}{f})\,(e).$$

Writing $\underset{\sim}{\psi}(X) = (X\,\underset{\sim}{f})(e)$, we have

$$\underset{\sim}{\psi}([X, Y]) = \pi(X)\,\underset{\sim}{\psi}(Y) - \pi(Y)\,\underset{\sim}{\psi}(X).$$

If we now consider V as a $\mathfrak{g}$ module and apply a Lemma due to Whitehead (Lemma 3, page 77, [6]) to the linear map $X \to \underset{\sim}{\psi}(X)$, we obtain

$$(\underset{\sim}{X}\ \underset{\sim}{f})\ (e) = \pi(X)\ \underset{\sim}{\alpha} \qquad \text{for all } X \epsilon \mathfrak{g} \tag{5.17}$$

for some $\underset{\sim}{\alpha} \epsilon V$. Hence

$$(X\ \underset{\sim}{f})\ (g) = g\ \pi(X)\ \underset{\sim}{\alpha}\ , \qquad g \epsilon G,\ X \epsilon \mathfrak{g}\ . \tag{5.18}$$

Thus for expt $X \epsilon N$, we obtain

$$\underset{\sim}{f}\ (\text{expt } X) = \sum_{0}^{\infty} \frac{t^n}{n!}\ (X^n\ \underset{\sim}{f})\ (e) =$$

$$\underset{\sim}{f}(e) + \sum_{1}^{\infty} \frac{t^n}{n!}\ \pi(X)^n\ \underset{\sim}{\alpha}\ . \tag{5.19}$$

Putting $g_1 = e$ in (5.14) we obtain $\underset{\sim}{f}(e) = 0$. Thus

$$\underset{\sim}{f}\ (\text{expt } X) = \{\exp t\ \pi(X)\}\ \underset{\sim}{\alpha} - \underset{\sim}{\alpha}$$

$$= (\exp t\ X)\ \underset{\sim}{\alpha} - \underset{\sim}{\alpha} \quad \text{if expt } X \epsilon N.$$

In other words

$$\underset{\sim}{f}(g) = g\ \underset{\sim}{\alpha} - \underset{\sim}{\alpha} \qquad \text{for all } g \epsilon N.$$

If for g_1 and g_2 the above equation holds, then

$$\underset{\sim}{f}(g_1\ g_2) = g_1\ \underset{\sim}{f}(g_2) + \underset{\sim}{f}(g_1)$$

$$= g_1\ (g_2\underset{\sim}{\alpha} - \underset{\sim}{\alpha}) + g_1\ \underset{\sim}{\alpha} - \underset{\sim}{\alpha}$$

$$= g_1 g_2\ \underset{\sim}{\alpha} - \underset{\sim}{\alpha}\ .$$

Since G is connected, N generates G. Thus for all $g \epsilon G$,

$$\underset{\sim}{f}(g) = g\,\underset{\sim}{\alpha} - \underset{\sim}{\alpha}\ .$$

This completes the proof of the lemma.

<u>Theorem 5.7</u> Let V be a real finite dimensional vector space and H be a compact or a connected semi simple analytic Lie group of endomorphisms of V. Let G be the semi-direct product of V and H. Let ω be a multiplier on G x G. Then there exists an equivalent multiplier ω_1 which is given by

$$\omega_1(n_1h_1,\ n_2h_2) = [\exp i\ B(n_1,\ h_1(n_2))]\,\delta(h_1,h_2)$$

for all $n_1, n_2 \epsilon V$, $h_1, h_2 \epsilon H$

where B is a skew symmetric bilinear form on V x V which is invariant under H and δ is a multiplier on H x H.

<u>Proof:</u> We shall first prove the theorem when H is connected and semi simple. By corollary 1 to Theorem 4.1 we may assume that ω is analytic in a neighbourhood of the identity. From Theorem 5.6 and its proof it is clear that ω can be assumed to be of the form

$$\omega(n_1h_1,\ n_2h_2) = \sigma(n_1,\ h_1(n_2))\ \delta(h_1,\ h_2)\ \psi(n_2,\ h_1)$$

where σ is a multiplier on V x V, δ is a multiplier on H x H and

ψ is a Borel function on V x H. Since equations (5.13) hold good with ω_1 replaced by ω it follows that σ, δ and ψ are analytic in a neighbourhood of the identity in the appropriate spaces.

By Theorem 5.4 and Lemma 4.1 we may assume that

$$\sigma(n_1, n_2) = a(n_1)\, a(n_2)\, a(n_1+n_2)^{-1} \exp i\, B(n_1, n_2) \tag{5.20}$$

where a(n) is analytic in a neighbourhood of the origin and B is a real skew symmetric bilinear functional. Further $|a(n)|=1$. Condition (i) of Theorem 5.6 implies that $\exp i\,[B(h(n_1),h(n_2)) - B(n_1,n_2)]$ is a symmetric function of n_1 and n_2 for every $h\varepsilon H$. Since $B(h(n_1),h(n_2)) - B(n_1,n_2)$ is skew symmetric in n_1 and n_2, this can happen if and only if

$$B(h(n_1), h(n_2)) = B(n_1, n_2) \qquad \text{for all } n_1,n_2\varepsilon V,\ h\varepsilon H. \tag{5.21}$$

This also implies that the function $\eta(n,h) = \psi(n,h)\, a(h(n)\, a(n)^{-1}$ on V x H satisfies the equation

$$\eta(n_1 n_2, h) = \eta(n_1, h)\, \eta(n_2, h) \qquad \text{for all } n_1,n_2\varepsilon V,\ h\varepsilon H.$$

Hence there exists a function $\underset{\sim}{f}(h)$ from H into V such that

$$\eta(n, h) = \exp i \,\langle \underset{\sim}{f}(h), n\rangle \quad \text{for all } n\varepsilon V,\ h\varepsilon H \tag{5.22}$$

where $\langle\cdot,\cdot\rangle$ denotes the standard Euclidean inner product (in some co-ordinate system) in V. Since by condition (ii) of Theorem 4.6,

$$\psi(n, h_1 h_2) = \psi(h_2(n), h_1)\, \psi(n, h_2) \qquad \text{for all } n \epsilon V,\ h_1, h_2 \epsilon H,$$

we have

$$\eta(n, h_1 h_2) = \eta(h_2(n), h_1)\, \eta(n, h_2) \qquad \text{for all } n \epsilon V,\ h_1, h_2 \epsilon H \tag{5.23}$$

From (5.22) and (5.23) we have

$$\langle \underset{\sim}{f}(h_1 h_2), n \rangle = \langle \underset{\sim}{f}(h_1), h_2(n) \rangle + \langle \underset{\sim}{f}(h_2), n \rangle \tag{5.24}$$

$$\text{for all } n \epsilon V,\ h_1, h_2 \epsilon H.$$

If we denote by h^* the adjoint of h relative to the inner product $\langle \cdot, \cdot \rangle$, we obtain

$$\underset{\sim}{f}(h_1 h_2) = h_2^* \underset{\sim}{f}(h_1) + \underset{\sim}{f}(h_2) \qquad \text{for all } h_1, h_2 \epsilon H.$$

Putting $\underset{\sim}{f}(h^{-1}) = \underset{\sim}{f}_1(h)$ we get

$$\underset{\sim}{f}_1(h_1 h_2) = h_1^{-1^*} \underset{\sim}{f}_1(h_2) + \underset{\sim}{f}_1(h_1) \qquad \text{for all } h_1, h_2 \epsilon H. \tag{5.25}$$

Since $g \to g^{-1^*}$ is an automorphism of the group of all non-singular endomorphisms of V, this takes the subgroup H into another subgroup H^* which is also connected and semi simple. If we define $\underset{\sim}{k}$ on H^* by the equation

$$\underset{\sim}{k}(h) = \underset{\sim 1}{f}(h^{*^{-1}}) \quad \text{for all } h \varepsilon H^*,$$

then we obtain from (5.25)

$$\underset{\sim}{k}(h_1 h_2) = h_1 \underset{\sim}{k}(h_2) + \underset{\sim}{k}(h_1) \quad \text{for all } h_1, h_2 \varepsilon H^*.$$

Since $\eta(n,h)$ is locally analytic it follows that $\underset{\sim}{k}(h)$ is also locally analytic. Hence by Lemma 5.2

$$\underset{\sim}{k}(h) = h \underset{\sim}{\alpha} - \underset{\sim}{\alpha} \quad \text{for all } h \varepsilon H^*$$

where $\underset{\sim}{\alpha}$ is a fixed vector in V. Thus

$$\underset{\sim}{f}(h) = h^* \underset{\sim}{\alpha} - \underset{\sim}{\alpha} \quad \text{for all } h \varepsilon H.$$

Thus from (5.22) we obtain

$$\begin{aligned} \eta(n, h) &= \exp i \langle h^* \underset{\sim}{\alpha} - \underset{\sim}{\alpha}, n \rangle \\ &= \exp i \langle \underset{\sim}{\alpha}, h(n) - n \rangle \end{aligned} \qquad (5.26)$$

Thus

$$\omega(n_1 h_1, n_2 h_2) = \delta(h_1, h_2)[\exp i\ B(n_1, h_1(n_2))]\, \eta(n_2, h_1) \frac{a(n_1)\ a(n_2)}{a(n_1 + h_1(n_2))} \qquad (5.27)$$

Putting $b(n\ h) = a(n)\ e^{-i\langle \underset{\sim}{\alpha}, n\rangle}$ we have

$$\omega(n_1 h_1, n_2 h_2) = \delta(h_1, h_2)[\exp i\ B(n_1, h_1(n_2))]\, b(n_1 h_1)\ b(n_2 h_2)\ b(n_1 h_1 n_2 h_2)^{-1}$$

Hence

$$\omega(n_1 h_1, n_2 h_2) \sim \delta(h_1, h_2) \exp i\ B(n_1, h_1(n_2)).$$

Equation (5.21) ensures the invariance of B under h. This completes the proof in the case when H is connected and semi simple. In the case of compact groups the only difference is in proving equation (5.26). This is an immediate consequence of Lemma 5.2 and hence the proof is complete.

Corollary 1 Suppose H is a compact group or a connected semi simple Lie group of endomorphisms in V which leaves a non singular symmetric bilinear form in V x V invariant. Let H be algebraically irreducible in the sense that every endomorphism commuting with all the elements of H is a scalar multiple of the identity. Let G be the semi direct product of V and H. Then every multiplier ω on G x G is equivalent to a multiplier ω_1 of the form

$$\omega_1(n_1 h_1, n_2 h_2) = \delta(h_1, h_2)$$

where δ is a multiplier on H x H. If in addition H is simply connected then every multiplier ω on G x G is trivial.

Proof: Because of Theorems 3.2, 5.1, 5.5 and 5.7 it is sufficient to prove that any skew symmetric bilinear form invariant under H is identically zero. Choose and fix a coordinate system in V and let A and B be the matrices of the given symmetric bilinear form and any invariant skew symmetric bilinear form. The invariance conditions imply that

$$\left.\begin{aligned} h\,A\,h^* &= A \\ h\,B\,h^* &= B \end{aligned}\right\} \quad \text{for all } h \in H.$$

Let $A^{-1}B = K$. A^{-1} exists since the symmetric form is non singular.

Then

$$AK = B = hBh^{*} = hAh^{*}h^{*^{-1}}Kh^{*}$$

$$= Ah^{*^{-1}}Kh^{*}.$$

Hence

$$K = h^{*^{-1}}Kh^{*} \qquad \text{for all } h \varepsilon H.$$

Thus h and K^{*} commute for all $h\varepsilon H$. Since H is algebraically irreducible, K is a scalar times the identity, i.e., $B = cA$. Since A is symmetric and B is skew symmetric this is impossible unless $c=0$. This completes the proof.

<u>Remark</u>: In Corollary 1 we may take for H the rotation group in R^3 and the Lorentz group in R^4.

REFERENCES

(1) V. Bargmann, On unitary ray representations of continuous groups, Ann. Math. Vol. 59, 1954, pp. 1-46.

(2) C. Chevalley, Theory of Lie Groups, Vol. 1, Princeton 1946.

(3) P. R. Halmos, Measure Theory, Van Nostrand, New York, 1950.

(4) S. Helgason, Differential Geometry and Symmetric Spaces, Academic Press, New York 1964.

(5) R. Hermann, Lie Groups for Physicists, Benjamin 1966.

(6) N. Jacobson, Lie Algebras, Interscience, New York, 1962

(7) G. W. Mackey, Borel structures in groups and their duals, Trans. Amer. Math. Soc. Vol. 85, 134-169, 1957

(8) G. W. Mackey, Unitary representations of group extensions I, Acta Mathematica Vol. 99, 1958, pp. 265-311.

(9) G. W. Mackey, Mathematical Foundations of Quantum Mechanics, Benjamin, New York, 1963.

(10) G. W. Mackey, Group representations, Mimeographed notes of lectures delivered at the Oxford Mathematics Institute 1966-67.

(11) K. R. Parthasarathy, Probability Measures on Metric Spaces, Academic Press, New York, 1967.

(12) D. J. Simms, Lie Groups and Quantum Mechanics, Lecture Notes in Mathematics, Springer-Verlag, 1968.

(13) V. S. Varadarajan, Geometry of Quantum Theory, Lecture Notes, Indian Statistical Institute, 1965.

(14) V. S. Varadarajan, Geometry of Quantum Theory, Vol. I, Van Nostrand, Princeton, 1968.

(15) J. Von Neumann, Mathematical Foundations of Quantum Mechanics Princeton University Press 1955 (English translation).

(16) E. P. Wigner, On the unitary representations of the Inhomogeneous Lorentz group, Ann. Math., Vol. 40, pp. 149-204, 1939.

(17) E. P. Wigner, Group Theory and its Applications to the Quantum Mechanics of Atomic Spectra, Academic Press, New York, 1959.

Offsetdruck: Julius Beltz, Weinheim/Bergstr.

Lecture Notes in Mathematics

Bisher erschienen/Already published

Vol. 1: J. Wermer, Seminar über Funktionen-Algebren. IV, 30 Seiten. 1964. DM 3,80 / 0.95

Vol. 2: A. Borel, Cohomologie des espaces localement compacts d'après J. Leray. IV, 93 pages. 1964. DM 9,– / $ 2.25

Vol. 3: J. F. Adams, Stable Homotopy Theory. 2nd. revised edition. IV, 78 pages. 1966. DM 7,80 / $ 1.95

Vol. 4: M. Arkowitz and C. R. Curjel, Groups of Homotopy Classes. 2nd. revised edition. IV, 36 pages. 1967. DM 4,80 / $ 1.20

Vol. 5: J.-P. Serre, Cohomologie Galoisienne. Troisième édition. VIII, 214 pages. 1965. DM 18,– / $ 4.50

Vol. 6: H. Hermes, Eine Termlogik mit Auswahloperator. IV, 42 Seiten. 1965. DM 5,80 / $ 1.45

Vol. 7: Ph. Tondeur, Introduction to Lie Groups and Transformation Groups. VIII, 176 pages. 1965. DM 13,50 / $ 3.40

Vol. 8: G. Fichera, Linear Elliptic Differential Systems and Eigenvalue Problems. IV, 176 pages. 1965. DM 13,50 / $ 3.40

Vol. 9: P. L. Ivănescu, Pseudo-Boolean Programming and Applications. IV, 50 pages. 1965. DM 4,80 / $ 1.20

Vol. 10: H. Lüneburg, Die Suzukigruppen und ihre Geometrien. VI, 111 Seiten. 1965. DM 8,– / $ 2.00

Vol. 11: J.-P. Serre, Algèbre Locale. Multiplicités. Rédigé par P. Gabriel. Seconde édition. VIII, 192 pages. 1965. DM 12,– / $ 3.00

Vol. 12: A. Dold, Halbexakte Homotopiefunktoren. II, 157 Seiten. 1966. DM 12,– / $ 3.00

Vol. 13: E. Thomas, Seminar on Fiber Spaces. IV, 45 pages. 1966. DM 4,80 / $ 1.20

Vol. 14: H. Werner, Vorlesung über Approximationstheorie. IV, 184 Seiten und 12 Seiten Anhang. 1966. DM 14,– / $ 3.50

Vol. 15: F. Oort, Commutative Group Schemes. VI, 133 pages. 1966. DM 9,80 / $ 2.45

Vol. 16: J. Pfanzagl and W. Pierlo, Compact Systems of Sets. IV, 48 pages. 1966. DM 5,80 / $ 1.45

Vol. 17: C. Müller, Spherical Harmonics. IV, 46 pages. 1966. DM 5,– / $ 1.25

Vol 18: H.-B. Brinkmann und D. Puppe, Kategorien und Funktoren. XII, 107 Seiten, 1966. DM 8,– / $ 2.00

Vol. 19: G. Stolzenberg, Volumes, Limits and Extensions of Analytic Varieties. IV, 45 pages. 1966. DM 5,40 / $ 1.35

Vol. 20: R. Hartshorne, Residues and Duality. VIII, 423 pages. 1966. DM 20,– / $ 5.00

Vol. 21: Seminar on Complex Multiplication. By A. Borel, S. Chowla, C. S. Herz, K. Iwasawa, J.-P. Serre. IV, 102 pages. 1966. DM 8,– / $ 2.00

Vol. 22: H. Bauer, Harmonische Räume und ihre Potentialtheorie. IV, 175 Seiten. 1966. DM 14,– / $ 3.50

Vol. 23: P. L. Ivănescu and S. Rudeanu, Pseudo-Boolean Methods for Bivalent Programming. 120 pages. 1966. DM 10,– / $ 2.50

Vol. 24: J. Lambek, Completions of Categories. IV, 69 pages. 1966. DM 6,80 / $ 1.70

Vol. 25: R. Narasimhan, Introduction to the Theory of Analytic Spaces. IV, 143 pages. 1966. DM 10,– / $ 2.50

Vol. 26: P.-A. Meyer, Processus de Markov. IV, 190 pages. 1967. DM 15,– / $ 3.75

Vol. 27: H. P. Künzi und S. T. Tan, Lineare Optimierung großer Systeme. VI, 121 Seiten. 1966. DM 12,– / $ 3.00

Vol. 28: P. E. Conner and E. E. Floyd, The Relation of Cobordism to K-Theories. VIII, 112 pages. 1966. DM 9,80 / $ 2.45

Vol. 29: K. Chandrasekharan, Einführung in die Analytische Zahlentheorie. VI, 199 Seiten. 1966. DM 16,80 / $ 4.20

Vol. 30: A. Frölicher and W. Bucher, Calculus in Vector Spaces without Norm. X, 146 pages. 1966. DM 12,– / $ 3.00

Vol. 31: Symposium on Probability Methods in Analysis. Chairman. D. A. Kappos. IV, 329 pages. 1967. DM 20,– / $ 5.00

Vol. 32: M. André, Méthode Simpliciale en Algèbre Homologique et Algèbre Commutative. IV, 122 pages. 1967. DM 12,– / $ 3.00

Vol. 33: G. I. Targonski, Seminar on Functional Operators and Equations. IV, 110 pages. 1967. DM 10,– / $ 2.50

Vol. 34: G. E. Bredon, Equivariant Cohomology Theories. VI, 64 pages. 1967. DM 6,80 / $ 1.70

Vol. 35: N. P. Bhatia and G. P. Szegö, Dynamical Systems. Stability Theory and Applications. VI, 416 pages. 1967. DM 24,– / $ 6.00

Vol. 36: A. Borel, Topics in the Homology Theory of Fibre Bundles. VI, 95 pages. 1967. DM 9,– / $ 2.25

Vol. 37: R. B. Jensen, Modelle der Mengenlehre. X, 176 Seiten. 1967. DM 14,– / $ 3.50

Vol. 38: R. Berger, R. Kiehl, E. Kunz und H.-J. Nastold, Differentialrechnung in der analytischen Geometrie IV, 134 Seiten. 1967. DM 12,– / $ 3.00

Vol. 39: Séminaire de Probabilités I. II, 189 pages. 1967. DM 14,– / $ 3.50

Vol. 40: J. Tits, Tabellen zu den einfachen Lie Gruppen und ihren Darstellungen. VI, 53 Seiten. 1967. DM 6.80 / $ 1.70

Vol. 41: A. Grothendieck, Local Cohomology. VI, 106 pages. 1967. DM 10.– / $ 2.50

Vol. 42: J. F. Berglund and K. H. Hofmann, Compact Semitopological Semigroups and Weakly Almost Periodic Functions. VI, 160 pages. 1967. DM 12,– / $ 3.00

Vol. 43: D. G. Quillen, Homotopical Algebra VI, 157 pages. 1967. DM 14,– / $ 3.50

Vol. 44: K. Urbanik, Lectures on Prediction Theory IV, 50 pages. 1967. DM 5,80 / $ 1.45

Vol. 45: A. Wilansky, Topics in Functional Analysis VI, 102 pages. 1967. DM 9,60 / $ 2.40

Vol. 46: P. E. Conner, Seminar on Periodic Maps IV, 116 pages. 1967. DM 10,60 / $ 2.65

Vol. 47: Reports of the Midwest Category Seminar I. IV, 181 pages. 1967. DM 14,80 / $ 3.70

Vol. 48: G. de Rham, S. Maumary et M. A. Kervaire, Torsion et Type Simple d'Homotopie. IV, 101 pages. 1967. DM 9,60 / $ 2.40

Vol. 49: C. Faith, Lectures on Injective Modules and Quotient Rings. XVI, 140 pages. 1967. DM 12,80 / $ 3.20

Vol. 50: L. Zalcman, Analytic Capacity and Rational Approximation, VI, 155 pages. 1968. DM 13.20 / $ 3.40

Vol. 51: Séminaire de Probabilités II. IV, 199 pages. 1968. DM 14,– / $ 3.50

Vol. 52: D. J. Simms, Lie Groups and Quantum Mechanics. IV, 90 pages. 1968. DM 8,– / $ 2.00

Vol. 53: J. Cerf, Sur les difféomorphismes de la sphère de dimension trois ($\Gamma_4 = O$). XII, 133 pages. 1968. DM 12,– / $ 3.00

Vol. 54: G. Shimura, Automorphic Functions and Number Theory VI, 69 pages. 1968. DM 8,– / $ 2.00

Vol. 55: D. Gromoll, W. Klingenberg und W. Meyer Riemannsche Geometrie im Großen VI, 287 Seiten. 1968. DM 20,– / $ 5.00

Bitte wenden / Continued

MIX
Papier aus verantwortungsvollen Quellen
Paper from responsible sources
FSC® C105338

If you have any concerns about our products,
you can contact us on
ProductSafety@springernature.com

In case Publisher is established outside the EU,
the EU authorized representative is:
Springer Nature Customer Service Center GmbH
Europaplatz 3, 69115 Heidelberg, Germany

Printed by Libri Plureos GmbH
in Hamburg, Germany